YOUR KNOWLEDGE HAS VALUE

Bibliographic information published by the German National Library:

The German National Library lists this publication in the National Bibliography; detailed bibliographic data are available on the Internet at http://dnb.dnb.de .

Imprint:

Copyright © 2008 GRIN Verlag, Open Publishing GmbH
Print and binding: Books on Demand GmbH, Norderstedt Germany
ISBN: 9783640600021

This book at GRIN:

http://www.grin.com/en/e-book/148172/information-communication-technologies

Ilir Hajdini

Information Communication Technologies

Information Communication Technologies' impact in International Companies' strategies, structures, productivity and environment pollution

GRIN Publishing

GRIN - Your knowledge has value

Since its foundation in 1998, GRIN has specialized in publishing academic texts by students, college teachers and other academics as e-book and printed book. The website www.grin.com is an ideal platform for presenting term papers, final papers, scientific essays, dissertations and specialist books.

Visit us on the internet:

http://www.grin.com/

http://www.facebook.com/grincom

http://www.twitter.com/grin_com

University of Sussex

Faculty of International Management
Brighton, Kings Road 51-53
United Kingdom
Ilir R. Hajdini
March 2008

Scientific paper in fulfillment of Master of Science degree in the subject of:

Managing Technology in Global Environment

titled:

Information Communication Technologies' impact in International Companies' strategies, structures, productivity and environment pollution

SPRU - SCIENCE AND TECHNOLOGY POLICY RESEARCH

Table of Contents

Acronyms and Abbreviations

Information (Communication) Technologies - is the computer and associated software that are used to receive, process, store, transmit and output data and information, including text, sound, graphics, and video. However, in a broader view, it consists of computers, networks, satellite communications, robotics, videotext, cable television, electronic mails, and other automated equipments. At the lowest level there are servers with operating systems, which have installed database and web serving software, which are than linked to other software and users via network infrastructures, who have their own hardware operating systems and software (Answers, 2008).

Enterprise Resource Planning (ERP) is a software which integrates a variety of business functions that has the potential of covering the entire value chain of activities, (manufacturing, logistics, distribution, inventory, invoicing) under a unified technological platform. (BPC, 2008)

Data Access and Analysis (DAA) is resource use system that provides easy access to company data, which permits analysis and identification of large volumes of data to make information available across functional boundaries and hierarchical levels in the organisation. It consists of Data Warehouses allowing access to Database Marketing, Statistical Sales Analysis tools and other departments of the Company (Poulymenakou at al, 2002).

Process Support and Improvement (PSI) is supposed to guide improvements in the content and context of the business operations in quality control, production, logistics, sales, distribution, customer care and services (Poulymenakou at al, 2002)

Information synergies (INS) Shows the effect that IT/ICT can play to increase the communication between-person or between-group individuals

3

within the organisation. In other words shows the performance gains that result when IT/ICT allows two or more individuals or subunits to pool their resources and cooperate/collaborate across roles, units or subunit boundaries (Dewett & Jones, 2001).

Information efficiencies (INE) is the ability to gather date and analyse them leading to cost and time savings that result when IT/ICT allows individual employees to perform their current tasks at a higher level, gain additional tasks, and expand their roles within the organization (Dewett & Jones, 2001).

Abstract

The essay will firstly describe some of the types of Information Communication Technologies (ICT) and its general impact on management, secondly elaborate the effect of ICT in International Companies' (IC) strategies, thirdly elaborate the effect of ICT in ICs' organizational structures, and on the way it will elaborate how can all this types of ICT as alternative resource-use systems impact the long-term wealth, well-being and productivity of ICs, and finally how ICT can help ICs operate in a more environmentally accepted way.

Introduction

The Business environment is characterized by increasing competition, sudden changes, and transformation in firms and market relationships due to innovations in products and services, market structures, and technology, as well as industries, and national boundaries (King & Sethi, 1999). In addition, the traditional protection between the firms and its environment, including time, people, and geography are reduced, leading to greater interdependence between the firms and its environments (King & Sethi, 1999). Similarly, governments are increasingly putting pressure in International Companies' strategies to perform in compliance with their environmental policies and standards.

Technology in general, and Information Technology (IT) or Information Communication Technology (ICT) in particular (which are often inextricably linked and, since it has become conventional to do so, the rest of this essay will mainly refer to them jointly as information communication technologies) are seen as critical forces in this transformation of competition, where International Companies (IC) have to rethink their strategies and

5

organizational structures in response to this new environment (King & Sethi, 1999).

However, still great majority of companies are not revolutionary, in their organizational strategies or structures. They are not truly aligned to the fast changing world outside. Indeed, a survey conducted in USA by the Gartner Group indicated that about 40% of ICT projects failed and that, on average, companies spent $1 million a year on unsuccessful projects, on top of wasted professional resources that could not be easily quantified or measured (Duh et al, 2006). The difficulty remains to fit Information Communication Technologies (ICTs) with the overall ICs' strategy and organizational structure. So, proper responding with the strategy and structure is a must for developing an effective ICT strategy, which should fit with the overall organization's vision, which if otherwise will cause a constant conflict (Colleen & Robert, 2006).

Thus, this essay will elaborate how ICs will be able to enhance the long term productivity by better integrating alternative resource-use systems, such as information communication technologies, to their strategies and organizational structure with acceptable environmental impacts.

Types of Information and Communication Technologies and its impact on management

This section will cover mainly ICT, which is achieved through technologies that are meant to improve the efficiency and effectiveness of business operations (Poulymenakou at al, 2002).

Let us start with Data Access and Analysis (DAA) as resource use system that provides easy access to company data, which allows analysis and identification of large volumes of data to make information available across functional boundaries and hierarchical levels in the organisation. It consists

6

of Data Warehouses permitting access to Database Marketing, Statistical Sales Analysis tools and other departments of the Company (Poulymenakou at al, 2002). Provided through DAA technologies, Management Decision Support (MDS) systems, on the other hand, support managers on decision making process with scenario evaluation, and monitoring of strategy implementation by handling largely unstructured, open-ended questions about unexpected future events (Poulymenakou at al, 2002). Similarly, Process Support and Improvement (PSI) is supposed to guide improvements in the content and context of the business operations in quality control, production, logistics, sales, distribution, customer care and services (Poulymenakou at al, 2002). Enterprise Resource Planning (ERP), on the other hand, is software which integrates a variety of business functions that has the potential of covering the entire value chain of activities, such as manufacturing, logistics, distribution, inventory and invoicing under a unified technological platform. (BPC, 2008).

The above description of the different impacts that resource use systems may have in IC, is cumulatively shown below in the figure 1, which hopefully will make the entire complex-picture much easier to understand:

Figure 1: information systems in relation to other business departments (Source[1])

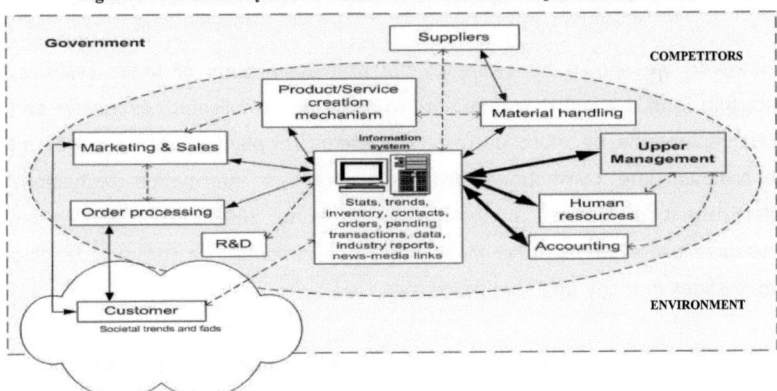

---

[1] The figure 1 is adopted from my team-work in the final year assignment studding at the University for Business and Technology, Prishtina, Kosova. More details about our group available at:
http://groups.yahoo.com/group/miscommunity/messages/101?viscount=100

The figure shows that information systems may integrate all parts of a business, such as manufacturing, human resources, planning, marketing, sales and far more. It evolves in response to complexity as more business departments need to be supported by a common information system. It tends to integrate all company departments and functions onto a single computer system, which will allow companies to standardize their operations, reduce data discrepancies between units, and implement best practices, as a result it shift its attention to external factors like their customers, partnering businesses, their competitors, government regulations possibly being friendlier to environment at the same time.

In other words it shows that ICT can be used to forecast demand for a product, order the necessary raw materials, establish production schedules, track inventory, allocate people, allocate costs, and project key financial measures. Similarly automating the supply chain is done by integrating ERP systems with other types of applications, such as e-commerce, and even with the computer networks of suppliers and customers (Answers, 2008). Another important benefit is that if appropriately used it may allow the company to replace complex computer applications with a single integrated system.

However, we should be aware of the draw-back pints of these systems, starting from the fact that they tend to be large, complicated, expensive and fast replaceable by more up-to-date systems. Implementation requires an enormous time commitment from a company's information technology department or outside professionals (Answers, 2008). This is because resource use systems affect most major departments in a company leading to changes in many business processes.

Thus, if carefully considered, analysed and adopted, all this alternative resource use systems, or types of ICT will be able to enhance the long-term productivity of the resource base and improve the long-term wealth and well-being of ICs by enabling the integration of, and access to, what has been called "organizational memory" and finally supporting managerial decision making (Poulymenakou at al, 2002). This will be shown in a deeper elaborated way in the following sections of this essay

The impact of ICT in international Companies' strategies

This section will mainly focus on ICTs enabling change through new opportunities associated with global strategies, transnational strategies or local strategies. And finally what ICs should take into account when implementing the alternative resource use systems

Companies operating globally face pressures for global strategy which means trying to minimize its unit costs (standardization) (Hill, 2007). On the other hand, localisation strategy means to be locally responsive, which is to differentiate products and marketing strategies to national preferences (customisation or differentiation) (Hill, 2007). However in today's competitive environment ICs have to lower prices (tend to use global strategy), and to respond to local wants and preferences at the same time (tend to use local strategy) which basically means to follow the so called transantional strategy (developed in the between of the two above strategies) (Daniels, 2007). The ICT gives the opportunity to the International Companies to either enhance the differentiation (localisation strategy), to lower costs (global strategy) (Poulymenakou at al, 2002), or to stay in between (transantional strategy).

ICT facilitating local (differentiation) strategy
Particularly, ICT allows a firm to achieve a differentiation advantage by securing relationships with customers through improved product/service quality and by enhancing its ability to quickly respond to market changes and requirements. Wal-Mart is a good example of how extensive use of ICT can provide a competitive advantage. In this particular case, the integration of the retailer's supply-chain software with DAA technologies ensures on-time

and efficient delivery of products to stores (Poulymenakou at al, 2002). Moreover, capitalising on Wal-Mart's and the largest personal and household producer Procter & Gamble (P&G) advanced information systems, a team of officers from both companies devised a way to increase process of current sales. Now, instead of sending sales representatives into stores, P&G examines the data itself, creates the order, and ships it as approved. The linkage allows P&G to custom, produce and ship to demand, substantially reducing inventories for both P&G and Wal-Mart as well (Corporate Watch, 2001)

ICT facilitating global (lower costs - standardisation) strategy

Furthermore, the ICT can enable cost advantages by playing a direct or indirect role in the cost of various activities in the value chain. For instance, P&G manufactures soap and detergent in a highly automated way. The company uses electronic data interchange (EDI) to optimize the purchasing process. Due to the high level of automation (Computers control production equipment and inventory management) the average plant has currently fewer than 20 employees (Hoovers, 2008). Similarly, Mexico's largest cement manufacturer Cemex sells ready-mixed cement that can survive for only about 90 minutes before solidifying, so precise delivery is important. But Cemex could not predict with total certainty what demand will be on a given day, week or month. To better mange unpredictable demand patterns, Cemex developed an integrative system of information technology, including track mounting global positioning systems, radio transmitters that allows Cemex to better control the production and distribution system, responding quickly to unanticipated changes in demand and reducing waste. The results are lower costs and superior customer service, which above many other factors allowed Cemex to position itself as the third largest cement company in the World (Hill 2007, p. 236).

ICT facilitating expansion strategies

Beyond differentiation and cost advantages, technology in general, and ICT in particular may also provide firms with new opportunities to expand to new markets and new types of activities (Poulymenakou at al, 2002). The combination of the availability of PCs and cheap pervasive fiber-optic cables all around the globe is causing a flexible worldwide information flow that allow workers anywhere, and any time, to interact. This makes possible to off-shoring of jobs as companies everywhere search the planet for the best talent at the best price (McFarlin & Sweeney, 2006). On other example supporting this argument is P&G, which use a global intranet system to network its research and development organisation allowing its 18,000 users to have access to more than five million pages of content. These resource use systems allow P&G to find people in the global R&D organisation with similar skills and interests and connect them with each other (Corporate Watch, 2001), which understandable leads to higher productivity and well being since the company will be able to gain best practices out of these different perspectives. Furthermore, the company also uses the Web as a new medium for marketing product and for enhancing customer service (Corporate Watch, 2001). They also use magazine advertising, coupons, and direct mails.

Thus, ICT utilization at this level influences managerial action through automated communication and collaboration (EDI, ERP, DAA, e-mail, Intranets) often crossing organizational boundaries (Internet) (Poulymenakou at al, 2002). In other words, it redefines opportunities and the choices ICs make to exploit those opportunities and establish new capabilities. As a result, ICs are able to gradually develop current business models and, in some cases, build new ones (Colleen & Robert, 2006).

However, the firm's information systems applications should be tied to the company's strategy. This is likely to affect most of the business functions, such as the affect it can have in people's job contents or spheres of influence and routines as well, which factors usually increase the employees' prone to resist the change. As a result, many companies find terrible difficulties to implement those information systems (Answers, 2008). In general, it is suggested that the failures to build new strategies in order to implement new resource use systems are most related to human and organizational factors, rather than technology-related factors (e.g., poor leadership, inadequate education and training, trying to maintain the status quo, and staff resistance) (Duh et al, 2006). In other words, adapting useful information systems for business involves many steps not typically associated with only controlling the flow of information. First of all, top level managers must take into consideration "people factors", secondly needs or desired outcomes, thirdly context or situational readiness, and finally the method or approach to managing change, see figure 2. (Isaksen & Tidd, 2006).

Figure 2 A model for implementing different resource use systems (a model for change)
Source: (Isaksen & Tidd, 2006)

People: If the ICs don't consider how their employees should respond and how they most likely will respond to their everyday situations, then the design will be flawed. An information system is not just new technology, but it should be a set of new work practices, enhanced by technological tools. Thus it requires new procedures, employee training, and both managerial and technical support (Answers, 2008).

Context: The ICs' Leaders or top managers should create the climate (context or situation) for change, they should explain to the employees the need to implement new systems, involving them in debates and decision making process, and finally being able to build trust within the organization, so when employees are asked to implement this new resource systems they should know the need for it, and they should be able to trust their leaders (Isaksen & Tidd, 2006).

Outcome: ICs should make employees well aware of the desired future vision as well as the core values, which many times are supposed to be held constant while the implementation of new systems takes place. The ICs should make very clear the degree of change that needs to be made, the relationship between the current and new desired outcome of the organisational strategy, and will there be new policies and procedures to apply new systems, or only improvements into existing ones (Isaksen & Tidd, 2006)..

Method: Technology policy regarding the adoption of new technologies is subject to managers' perceptions towards the direction and pace of their companies, industries and market changes in general. In other words, there is no single method that can be supposed as 'magic bullet'. All methods have their own costs and benefits differing in their ability to fit various circumstances (Isaksen & Tidd, 2006). Thus Irrespective of how accurately these perceptions reflect objective reality, the fact remains that decision to adopt or not new technologies are moderated by the particular understanding the firm develops of its market circumstances (Poulymenakou at al, 2002).

Nevertheless, there is a strong belief supported with lots of examples and arguments by Tidd and Isaksen that incremental implementation of new systems will result almost always to be much more effective, which above many other factors reduces people resistances (Isaksen & Tidd, 2006). This essay supports this view, arguing that different types of resource use systems did not appear to be used in all companies at the same time, or at once. Well, they have influenced the strategies and structures, but this has been in relation with other factors.

In addition, the above ICT systems were used step by step in relation with market needs to be more efficient and effective, and to, if properly implemented, increase productivity and well being of ICs by enabling the integration of, and access to, what has been called "organizational memory" and finally supporting managerial decision making (Poulymenakou at al, 2002). Thus, ICT does facilitate the day to day business operations by enabling standardisation or customisation of their offerings increasing above many other opportunities competitiveness as well.

To sum up, as new technology becomes very strong and important part of the organization facilitating the day to day operations, gradual learning occurs that increases the awareness of different opportunities for action, which in turn should translate into incremental changes in strategic goals (Poulymenakou at al, 2002).

The impact of ICT in international Companies' organisational structures

Organisational structures have a significant importance in achieving strategic goals, in fact it is considered to be as a tool to the achievement of strategic goals. More specifically in the journal of Management & Information Poulymenakou (2002) considers organisational structures as:

"the context for strategic choices to be formulated and it also provides the vehicle through which these choices are effectively implemented" (Poulymenakou at al, 2002).

In general, structure can be defined as mechanistic characterized by a combination of a tall vertical hierarchy with high degrees of formalization and centralization, or on the other hand, organic structure characterized by fewer managerial units and more decentralized (Poulymenakou at al, 2002). The differences between both are shown in the table bellow:

Table 1: Mechanistic (formalised) and Organic (decentralised) organisational structures

Mechanistic	Organic
Individual specialization: Employees work separately and specialize in one task	Joint Specialization: Employees work together and coordinate tasks
Simple integrating mechanisms: Hierarchy of authority well-defined	Complex integrating mechanisms: task forces and teams are primary integrating mechanisms
Centralization: Decision-making kept as high as possible. Most communication is vertical.	Decentralization: Authority to control tasks is delegated. Most communication lateral
Standardization: Extensive use made of rules & Standard Operating Procedures	Mutual Adjustment: Face-to-face contact for coordination. Work process tends to be unpredictable
Much written communication	Much verbal communication
Informal status in org based on size of empire	Informal status based on perceived brilliance
Organization is a network of positions, corresponding to tasks. Typically each person corresponds to one task	Organization is network of persons or teams. People work in different capacities simultaneously and over time

Source[2]

ICT facilitating mechanistic structure

As it can also be seen in the table mechanistic structure or formalization can be achieved through the use of rules and standard operating procedures and through the development of common and shared norms and values (Dewett & Jones, 2001). Basically, formalization speaks to the desire for less

---

[2] **(Analytic-technologies, 2008 online)**

15

ambiguity and more efficiency goals, which ICT may be particularly suited to address. This can be said, since ICT facilitates the recording and re-checking of information about ICs' events and activities making the control of behaviours and processes through formalization more effective or viable (Dewett & Jones, 2001). It offers the ability to reduce the negative effects of formalization leading to information efficiencies (INE) by increasing efficiency, reducing search times, interruptions in workflow, and the administrative cost of formalization (Dewett & Jones, 2001). Support for this perspective is provided by Groth (1999) who analyzed the way ICT affected the development of the Boeing 777, which used organic structure (formal decision making). Boeing 777 was the first aircraft which was completely designed using alternative ICT systems that coordinated the activities of over 5000 people at over twenty sites in two different countries. Boeing's new resource use system managed thousands of drawings and documents which were studied, evaluated, updated, and changed constantly. Subsequent use of this system has allowed Boeing to create custom versions of the aircraft in 18 months compared to the traditional average of 52 months (Dewett & Jones 2001, p.18), therefore enabling the company to save considerable amount of costs and time.

ICT facilitating organic structure

On the other hand, organic structure has increasingly been used because of increased domestic and global competition throughout the 1990s. Due to it, many firms today have begun to move strategic decision making lower in the organization to take advantage of specialized workers who have gained more accurate and timely local information (Dewett & Jones, 2001). ICT directly improves such attempts in two major ways. First, they result in INE because they increase local information by enriching it with more intimate knowledge of consumer and market trends and opportunities (Dewett & Jones, 2001). For example, ICT in customer support centres has become a widespread means of increasing efficiency directed at solving customer problems via the internet. Secondly ICT can produce information synergies (INS) because they

16

facilitate increased communication and coordination between central planners, decentralized decision makers and upper management so that local action do not necessarily become more fragmented with respect to International Companies' goals as decision making authority moves lower in the hierarchy, but may actually become better aligned (Dewett & Jones, 2001).

A more fundamental question is whether or not ICT will lead to centralization or decentralization?

In terms of centralization, by enabling managers to get information more quickly and accurately, management information systems reduce uncertainty and lead managers to make decisions that they otherwise may not have made (Dewett & Jones, 2001). In contrast, decentralization through other forms of ICT (e.g., electronic bulletin boards and discussion groups) enable lower and middle level managers to stay better informed about the IC's overall situation and about the nature of current problems and issues (Dewett & Jones, 2001). The literature suggests that ICT can thus facilitate and enable both centralization and decentralization. Researchers seem to agree that the use of ICT allows organizations to place decision making authority across a greater range of hierarchical levels without sacrificing decision quality or decision efficiency combined the notions of centralization and decentralization in what is termed a transantional organization (It operates with transnational strategy) (Dewett & Jones, 2001). The effect is to move authority towards that part of the organization where the pertinent data is to be used in an effective way to make informed decisions, supporting information efficiency (Dewett & Jones, 2001).

The extra literature generally predicts a flattening of the organizational structure because of changes in ICT. Supporters of this view describe that flattening of managerial hierarchy is happening, because the need for information intermediaries is reduced (Poulymenakou at al, 2002). According to this, the role of middle managers as information supervisors tends to be seen as not necessary any more, since ICT permits top management to communicate desired information without middle management. The decision-making process will, therefore, involve fewer levels of the organization because access to information is greater and easier with ICT (Poulymenakou at al, 2002). However, there are objections to these opinions, arguing that the role of ICT is not to replace middle-level professionals, but instead, to leverage their work (Poulymenakou at al, 2002).

To conclude, ICTs by increasing the level of formalization or allowing "controlled decentralization" can act as a substitute for the control provided by a hierarchy in ICs (Poulymenakou at al, 2002). In addition, since ICT provides lower-level employees with more freedom to coordinate their actions and tasks, this result in INS as employees can experiment and find better ways of performing their tasks and responsibilities. Support for this point of view is found in the increasing number of flat, empowered, organizational structures with virtual organizations being an extreme case of low-cost companies that has begun to materialize (Poulymenakou at al, 2002). Therefore, ICT reduces the flow of information across vertical levels and knowledge flows more freely cross-functionally giving more opportunities that can be supported with more flexible decision making authority across levels of the hierarchy in the company (Poulymenakou at al, 2002).

How does ICT 'reduce' negative environmental impacts?

Austin and Macauley (2001) argue that technology is a double-edged sword-one capable both of doing and undoing damage to environmental quality. They continued by saying that it took nearly three generations before the first concerted efforts were made to bring pollution under control, but once begun, progress has been real (Austin & Macauley, 2001). Moreover, the revolution in ICT promises environmental improvements almost as great as those of the industrial revolution itself. As a matter of fact ICT is substituting the traditional logistic networks as the highways by which service, goods are transferred (Matthews, 2004). For instance, the online shopping replaces visits to the mall or takes place in addition to trips to the dentist and dry cleaners, which shows the Internet's impact on reducing energy consumption and environmental pollution. In addition, ERP, DAA and other ICT types are leading to digital data storage, manipulation, and communication with high-speed and high-bandwidth connectivity between our homes and offices allowing us to telecommute (Street, 2007). Therefore, ICT is helping organizations to become greener though adoption of practices like flexible working and increased efficiencies in business processes. For example, Salford and Bradford have both introduced full time home based working for benefits service staff. With the right electronic document and records management (EDRM) and workflow monitoring systems and administrative processes in place, both Councils have reported significant increases in staff productivity, retention and well being, as well as rapid payback on investment (Street, 2007). Therefore, this resource use systems are having considerable environmental implications (Street, 2007). Another indirect positive implication is Internet's role to expanded the public's access to and awareness of detailed environmental information informing decision-makers on matters of environmental concerns

Similarly, organisations introducing mobile working have made similar gains. Cumbria County Council substantially reduced the number of journeys Social Services staff made to the office when the team was given Tablet PCs to record home-based interviews with clients (Street, 2007). Replacing face-to-face meetings with digital equivalents can be similarly cost-effective: BT claims to have saved £400m since 2001 by using video conferencing (Street, 2007).

Another important benefit is that it allows the company to replace complex computer applications with a single, integrated system. Not only this, regarding the environment pollution, ICT services are known to be effective in reducing CO2 emissions by helping to minimize the movement of people and goods and the use of paper and other office supplies (NTT, 2006). The online shopping website, ebay, that can be used to by goods online, will reduce the use of warehouses as well as trips to let say the bookstores, or work related journeys. This view is supported by Austin & Macauley (2001) as well, who consider that ICTs' key importance is its role to reduce carbon emissions by minimizing work-related journeys.

Nevertheless, there are many pints behind those positive aspects of ICT use, starting from the energy use of transportation for delivering the ordered products (books for instance), and the packaging needed for each item ordered, so instead of shipping retail stores a box packed with 15 copies of a book, clients are shipped single boxes with a single book, trucks may carry fewer ordered material (customised), thus adding unnecessary packaging material, and more often delivery journeys leading to higher energy use, and carbon emission (Matthews, 2004). Secondly, teleconferencing or teleworking has given the opportunity to employees for additional shopping trips or other trips leading to again more use of energy and higher carbon emissions. This is known as a 'rebound effect' (Matthews, 2004). Thirdly, recycling of all this hardware and software, and the energy the currently consume is yet to be improved (Hilly et al, 2006). Finally, time has shown that people may feel unhappy working only through means of computers therefore it may be unproductive, since face to face meetings (conversations) are much more understandable and motivating.

On the other hand we can argue that ICT is still to be improved, for example, radio frequency identification can be used to produce and pack containers more efficiently and assess transportation needs. Global positioning systems is used to find a location anywhere on the planet, which is similar to navigators that modern car and telephone companies (Toyota, Nokia) are currently using, and geographic information systems that facilitates linking information with geographic locations (Matthews, 2004). This all alternative resource use systems can improve energy use, waste and carbon emissions by better managing production, packaging, and manage telecommunication and transportation.

To summaries, virtualization of products, digitization of information, dematerialization of transport, diminishing of warehouses/office spaces and shortening of supply chains are all, at first glance, positive impacts (Matthews, 2004). Cases of CDs to mp3s, catalogues to websites, flights to teleconferencing (the average translantic business flight uses 80,000 to 100,000 lb of fossil fuel, which could easily be avoided by the use of teleconferencing) are very good arguments to support ICT positive impact in environmental concerns (Yi & Thomas, 2006), Thereafter its benefits lie in the second order effects via increased efficiency, transparency, and speed of transactions. However there are many points to be considered from improvement of ICTs to better manage the production and transportation process till to the consideration of concerns related to people who need some physical contact not just video conferencing and teleworking

Conclusion

Presumably, all companies nowadays have seen the importance of incorporating and properly using the technology in general, or ICT in particular. However, not a lot has been done to understand the linkage between the ICT, strategy, organisational structure and its effects in overall productivity. Thus, this essay may have brought you to the conclusion that ICTs can support decision making by covering the entire value chain of activities under a unified technological platform, guiding improvements in the content and context of the business processes in which they are embedded, permitting analysis and identification of relations "hidden" in large volumes of data to make information available to a wide set of actors across functional boundaries and hierarchical levels thus allowing employees or subunits to pool their resources in cooperation and collaboration. More practically, cost and time savings that result when ICT allows individual employees to perform their current tasks at a higher level, assume additional tasks, and expand their roles in the organization due to advances in the ability to gather and analyze data, thus finally facilitating the achievement of the strategic goals, regardless their differentiation or standardisation nature of the strategy.

On the other hand, it was indicated that investment in ICT is no guarantee of business success, since it can fast replaceable with a newer version and it is very complicated to transfer the knowledge necessary to employees in order for them to properly use it, thus a focus on increasing market share or any other strategy may lead to better performance for many companies. Further complications are caused since the same resource use systems that have the potential of eliminating the need for people to do dangerous work improving working conditions and overall performance of operations, can carry very threats starting from, de-motivation, resistance to change, job displacements in one hand, and higher indirect carbon emissions, energy consumption by unnecessary customised product offers and transportation directions on the other hand.

To summarise, the green and other benefits of ICT in general requires a highly flexible working strategy that provides a powerful business case for its adoption. The key lies in careful preparation, so that the changes needed in information and cultural management should be carefully aligned with strategic goals of the company and consequently its organisational structure and business processes in general. Therefore is not the matter of radical changes, but of a careful consideration to build a flexible strategy that leaves the door open for businesses to be redefined, corporate culture to change, for ICTs use and its relation to environmental concerns to be revisited.

Bibliography

Hill, M. (2007): *International Business*: *Competing in the Global Marketplace,* 6th edition, ch.7, 12, 13, 14, pp. 229-430, New York, McGraw-Hill/Irwin.

Daniels, J. Radebaugh, L. Sullivan, P. (2007): *International Business: Environments and Operations,* 11th edition, ch.9, pp. 392-399, New Jersey, Prentice-Hall, Inc.

McFarlin, D., Sweeney, P. (2006) International Management: Strategic Opportunities and Cultural Challenges (3rd Edition), Boston, Mass.: Houghton Mifflin Company, Chapter 1, Page 4 -5

Isaksen, S., Tidd, J. (2006): Meeting the Innovation: Leadership for Transformation, 4th edition, ch.5, pp. 119-152, England, John Wiley & Sons, Ltd

King, W., Sethi, V. (1999) An empirical assessment of the organization of transnational information systems. Journal of Management Information Systems, [online], Vol. 15, Iss. 4; pg. 7, 22, Available at: http://proquest.umi.com/pqdlink?index=0&did=42071833&SrchMode=1&sid =4&Fmt=4&VInst=PROD&VType=PQD&RQT=309&VName=PQD&TS=120346 6275&clientId=5646 [Last Accessed: 27 – Feb – 08].

Austin, D., *Macauley,* M. (2001) Cutting through environmental issues. Journal of Management Information Systems, [online], Vol. 19, Iss. 1; pg. 24, 4, Available at: http://proquest.umi.com/pqdweb?index=7&did=66379303&SrchMode=1&sid =2&Fmt=4&VInst=PROD&VType=PQD&RQT=309&VName=PQD&TS=120391 5403&clientId=5646 [Last Accessed: 27 – Feb – 08].

Dewett, T., Jones, G. (2001) The role of information technology in the organization: a review, model, and assessment. Journal of Management, pg 339-351, Available at: file:///F:/New%20Folder/The%20role%20of%20information%20technology %20in%20the%20organization%20a%20review,%20model,%20and%20ass essment%20--%20Dewett%20and%20Jones%2027%20(3)%20313%20-- %20Journal%20of%20Management.htm [Last Accessed: 27 – Feb – 08].

Poulymenakou, A., Prastacos, G., Spanos Y. (2002) The relationship between information and communication technologies adoption and management. Information & Management, [online], Volume 39, Issue 8, Pages 659-675 Available at: http://www.sciencedirect.com/science?_ob=ArticleURL&_udi=B6VD0- 44X0590-

1& user=128860& coverDate=09%2F30%2F2002& alid=692820877& rdoc
=39& fmt=full& orig=search& cdi=5968& sort=d& docanchor=&view=c&
ct=107& acct=C000010638& version=1& urlVersion=0& userid=128860&
md5=2a8ac7844a7bbb5d5e1ca9e31f0b506e&artImgPref=F [Last Accessed:
27 – Feb – 08].

Colleen, Sh., Robert, B. (2007) Developing and Implementing a Strategy for
Technology Deployment. Information Management Journal, Available at:
http://findarticles.com/p/articles/mi_qa3937/is_200607/ai_n17176090 [Last
Accessed: 27 – Feb – 08].

Street, T. (2007) Green ICT – taking the strategic approach. Socitm
Consulting, Available at: http://www.socitm.gov.uk/NR/rdonlyres/282D6A8F-
56E1-4F7D-A566-
C430A816BA1D/0/10130SocitmConsultingGreenICTbriefing0907.pdf [Last
Accessed: 27 – Feb – 08].

Duha, R., Chowb, Ch. & Chenc, H. (2006) Strategy, next term IT applications
for planning and control, and firm performance: The impact of impediments
to IT implementation. Information & Management, [Online], Volume 43,
Issue 8, Pages 939-949 Available at:
file:///C:/Documents%20and%20Settings/Ilir%20%20Hajdini/My%20Docum
ents/Essay5000/science.htm [Last Accessed: 28 – Feb – 08].

Hung, Sh. & Tang, R. (2007) Factors affecting the choice of technology
acquisition mode: An empirical analysis of the electronic firms of Japan,
Korea and Taiwan. Technovation , Available at:
http://www.sciencedirect.com/science?_ob=ArticleURL&_udi=B6V8B-
4R7CYGT-
1& user=128860& coverDate=11%2F28%2F2007& alid=692820877& rdoc
=8& fmt=full& orig=search& cdi=5866& sort=d& docanchor=&view=c& ct
=107& acct=C000010638& version=1& urlVersion=0& userid=128860&md
5=c0c68783389cf3142063634c557c32f7 [Last Accessed: 29 – Feb – 08].

Hiltya, L., Arnfalkb, P., Erdmannc, L., Goodmand, L., Lehmanna, M., Wägera,
P. (2006) *Environmental Modelling & Software: The relevance of information
and communication technologies for environmental sustainability – A
prospective simulation study,* [Online], Volume 21, Issue 11, November
2006, Pages 1618-1629, Available at:
http://www.sciencedirect.com/science?_ob=ArticleURL&_udi=B6VHC-
4KCGJ5S-
1& user=128860& coverDate=11%2F30%2F2006& alid=704486392& rdoc
=4& fmt=full& orig=search& cdi=6063& sort=d& docanchor=&view=c& ct
=10& acct=C000010638& version=1& urlVersion=0& userid=128860&md5
=2cc9d04b1f4932af06c97bd9f13000c5 [Last Accessed: 05 – March – 08].

Yi, L., Thomasa, H. (2007) Environment International: A review of research on the environmental impact of e-business and ICT, [Online], Volume 33, Issue 6, August 2007, Pages 841-849, Available at: http://www.sciencedirect.com/science?_ob=ArticleURL&_udi=B6V7X-4NP3P4G-1&_user=128860&_origUdi=B6V9G-46H7TWJ-8&_fmt=high&_coverDate=08%2F31%2F2007&_rdoc=1&_orig=article&_acct=C000010638&_version=1&_urlVersion=0&_userid=128860&md5=b99672b33f91ca5c6519c0694fac193f [Last Accessed: 10 – March – 08].

Matthews, S. (2004) Information Technology and the Environment: Reflections on Current Research and Understanding, [Online], Available at: http://www.environmentalfutures.org/Images/nsfitwhitepaper.pdf [Last Accessed: 11 – March – 08].

BBC, 2007, Basic Skills and ICT. [Online]. Available at: http://www.bbc.co.uk/skillswise/tutors/expertcolumn/ict/ , [Last Accessed: 27 – Feb – 08].

Answers.com, 'Enterprise Resource Planning (ERP)'. [Online]. Available at: http://www.answers.com/topic/enterprise-resource-planning?cat=technology, [Last Accessed: 021-Feb-08].

Analytictech, [Online]. Available at: http://www.analytictech.com/mb021/organic_vs_mechanistic_structure.htm [Last Accessed: 021-Feb-08].

Hoovers, Soap and Detergent Manufacturer, [Online]. Available at: http://premium.hoovers.com/subscribe/ind/fr/profile/basic.xhtml?ID=223, [Last Accessed: 25 – Feb – 08].

Corporate Watch, 2007, Influence / lobbying groups. [Online]. Available at: http://www.bbc.co.uk/skillswise/tutors/expertcolumn/ict/ , [Last Accessed: 23 – Feb – 08].

NTT Group, 2006, Developing Methods for Calculating and Evaluating the Effectiveness of ICT in Reducing Environmental Impacts. [Online]. Available at: http://www.ntt.co.jp/csr_e/2006report/ecology/07.html [Last Accessed: 23 – Feb – 08].

BPC Articles and Glossary, 2008, Enterprise Resource Planning Software [Online]. Available at: http://www.bestpricecomputers.co.uk/glossary/enterprise-resource-planning.htm [Last Accessed: 24 – Feb – 08].